RAIN

WHERE DO YOU COME FROM?

By Francesca Grazzini

Illustrated by Chiara Carrer

Translated by Talia Wise

A CURIOUS NELL BOOK

KM Kane/Miller Book Publishers

Brooklyn, New York & La Jolla, California

Oh no, it's raining cats and dogs!
Good morning rain drops.
Do you know you all look alike?

Of course, there's even an old saying,
"to be as alike as two rain drops."

I like the water very much. I love swimming. I was born in the water, and as a tadpole, I lived only in the water. But, after awhile, I was able to breathe air and could live on land as well.

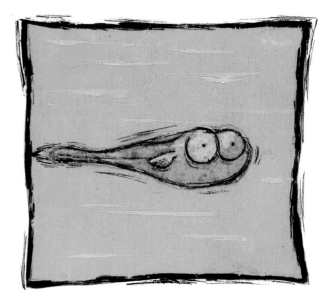

This is me as an egg alongside my brothers and sisters. And this frog is my mom.

Here I am as a . . . tadpole!

Some people believe that the first animals ever to live on earth lived in water, and only after they learned to breathe air did they leave the water to populate the rest of the planet.

And here I am after I grew legs!

Here I am as a frog when I could finally leave the water.

Of course there are still many animals and plants that live only in water . . .

GOLDFISH

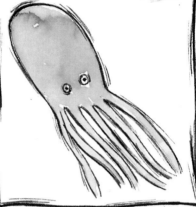

OCTOPUS

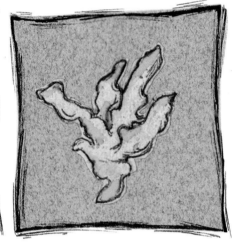

SEAWEED

OYSTER

TROUT

SEA URCHIN

Do you recognize any of them?

SWORDFISH

HERMIT CRAB

BLOWFISH

WATER LILY

WHALE

SHRIMP

What about places where there isn't much water?

They are called deserts.
That's where cactus plants grow.
Do you know why they look so fat?
It's because they fill themselves
up with water when it rains.
They are able to store this water
inside themselves for months and
months and then use it as
needed.

Cacti also have thorns. These thorns protect them from animals who would otherwise try to drink all of their stored water!

Camels who also live in the desert have a storage place too! Many people think they store water in their humps, but that's not true. (Actually, it is the fat produced from food that is stored in their humps.) The water they drink is stored in pouches in their stomachs and released as they need it.

Goodbye frog! I really must go.
I'm on my way to the sea!

Can't you take me
with you?

I don't think that's a good idea!
You like me now because I'm freshwater
and I taste good, and you can drink me.
But in the sea, I become saltwater, and
then you won't want to drink me.

SONG OF THE WATER DROP

From up in the clouds
To deep in the sea
I travel afar to where I must be . . .
But I'm sorry to say
You can't travel with me
Have you ever seen a frog down by the sea?

Don't be sad my friend!
Sooner or later I'll come back to find
you, for my journey is like a big circle . . .
I always return to the place I begin.
So, if I fall on the mountains as rain,
I join the stream that goes down the
mountain, and that stream joins a river,
and then that river ends up at the sea!

SEA

And then what happens?

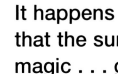

It happens . . . it happens
that the sun performs
magic . . . great magic!

What does the sun do?

The sun changes me.
When I'm water in the sea on
a beautiful sunny day, the sun heats
me and makes me evaporate.

This means I begin to change into teeny bits of water called vapor. Vapor is so light that it moves up high into the sky where it's much colder. The coldness makes the tiny drops of water, or vapor, move closer together. Together, many, many tiny drops of water make clouds.

Do clouds change their form and move around because of the wind?

Yes, exactly!
Clouds, pushed by the wind,
circulate . . . you can see this movement
of clouds on television on the . . . meto
. . . meteorological . . .

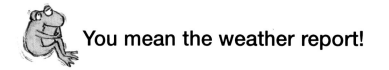

You mean the weather report!

In the clouds many tiny drops of water come together to form much bigger water drops, and when they become big enough and heavy enough, they fall as rain or snow.

So you see, I will return!
I will come down from the clouds!
I will begin my trip all over again
. . . but there's more magic!

If I fall where it is cool, I will be . . .

. . . rain drops!

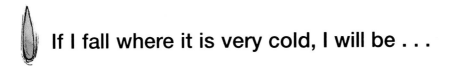 If I fall where it is very cold, I will be . . .

. . . snowflakes!

I see! It's like boiling water in a pot. When the water boils, we can see the water evaporate and turn into steam, or vapor. And if I put a cold lid on the pot, the vapor will move closer together on the lid and change into little drops of water!

Yes, that's it!
But now I really must go!
Goodbye! Or better yet . . .
see you again!

WATER GAME

If I am dew . . .

I help the rose.

If I am fog . . .

what a bore!

If I am snow . . .

bring your overcoat!

If I am hail . . .

you have to run,
if you can!

If I am a cloud . . .

I travel the sky.

If I am rain . . .

open your umbrella
wide!

In the stream . . .

I quietly murmur.

From the spring . . .

you can drink me.

In the pond . . .

so many mosquitoes!

In the river . . .

you can sail.

As a waterfall . . .

I am unrestrained!

And in the lake . . .

I blissfully sleep.